Cambridge Elements ≡

Elements in Geochemical Tracers in Earth System Science
edited by
Timothy Lyons
University of California
Alexandra Turchyn
University of Cambridge
Chris Reinhard
Georgia Institute of Technology

PELAGIC BARITE

A Tracer of Ocean Productivity and a Recorder of Isotopic Compositions of Seawater S, O, Sr, Ca, and Ba

Weiqi Yao
University of Toronto

Elizabeth Griffith
Ohio State University

Adina Paytan
University of California, Santa Cruz

CAMBRIDGE
UNIVERSITY PRESS

CAMBRIDGE
UNIVERSITY PRESS

University Printing House, Cambridge CB2 8BS, United Kingdom

One Liberty Plaza, 20th Floor, New York, NY 10006, USA

477 Williamstown Road, Port Melbourne, VIC 3207, Australia

314–321, 3rd Floor, Plot 3, Splendor Forum, Jasola District Centre,
New Delhi – 110025, India

79 Anson Road, #06–04/06, Singapore 079906

Cambridge University Press is part of the University of Cambridge.

It furthers the University's mission by disseminating knowledge in the pursuit of
education, learning, and research at the highest international levels of excellence.

www.cambridge.org
Information on this title: www.cambridge.org/9781108810692
DOI: 10.1017/9781108847162

© Weiqi Yao, Elizabeth Griffith, and Adina Paytan 2020

First published 2020

A catalogue record for this publication is available from the British Library.

ISBN 978-1-108-81069-2 Paperback
ISSN 2515-7027 (online)
ISSN 2515-6454 (print)

Pelagic Barite

A Tracer of Ocean Productivity and a Recorder of Isotopic Compositions of Seawater S, O, Sr, Ca, and Ba

Elements in Geochemical Tracers in Earth System Science

DOI: 10.1017/9781108847162
First published online: December 2020

Weiqi Yao
University of Toronto

Elizabeth Griffith
Ohio State University

Adina Paytan
University of California, Santa Cruz

Author for correspondence: Adina Paytan, apaytan@ucsc.edu

Abstract: Reconstruction of ocean paleoproductivity and paleochemistry is paramount to understanding global biogeochemical cycles such as the carbon, oxygen, and sulfur cycles and the responses of these cycles to changes in climate and tectonics. Paleo-reconstruction involves the application of various tracers that record seawater compositions, which in turn may be used to infer oceanic processes. Several important tracers are incorporated into pelagic barite, an authigenic mineral that forms in the water column. Here we summarize the utility of pelagic barite for the reconstruction of export production and as a recorder of seawater S, O, Sr, Ca, and Ba.

Keywords: biogeochemical cycles, geochemistry, marine barite, ocean productivity, seawater chemistry, stable isotope

ISBNs: 9781108810692 (PB), 9781108847162 (OC)
ISSNs: 2515-7027 (online), 2515-6454 (print)

Contents

1 Pelagic Barite Accumulation Rates: A Proxy for Export Production

Barite ($BaSO_4$) precipitates in the ocean at several settings where supersaturation with respect to barite is achieved. These conditions occur in microenvironments in the oceanic water column (pelagic barite), at volcanic hydrothermal settings (hydrothermal barite) or cold seeps (cold-seep barite), or in porewaters beneath the sediment–water interface (diagenetic barite). Pelagic barite crystals that are found in the water column or marine sediments are elliptical or euhedral, ranging in size from 1 to 5 μm (Figure 1). Pelagic barite can be found throughout the oceanic water column associated with sinking particulate matter and in marine sediments underlying areas of high biological productivity (Dehairs et al., 1980, 1991, 2000; Bishop, 1988; Dymond et al., 1992; Francois et al., 1995; Dymond and Collier, 1996; Paytan et al., 1996a; Paytan and Griffith, 2007). Precipitation of barite especially at water depth of 200–1500 meters and its preservation throughout the water column and in marine sediments has been a long-standing paradox, owing to the undersaturation of the world ocean with respect to barite (Monnin et al., 1999; Rushdi et al., 2000; Monnin and Cividini, 2006). Barite formation in undersaturated seawater and thus requires some biological or abiotic mechanism to create supersaturated microenvironments in which barite can precipitate. Although some marine organisms can form barite within their cells (Gooday and Nott, 1982; Finlay et al., 1983), their occurrence appears too sparse to account for the ubiquitous presence of barite in seawater. Similarly, the presence of Acantharea – organisms containing considerable quantities of Ba that precipitate $SrSO_4$ (celestite) (e.g., Bernstein and Byrne, 2004) – is not necessary for driving significant Ba removal from seawater (Esser and Volpe, 2002) or for barite precipitation (Ganeshram et al., 2003). Instead, barite micro-crystals are thought to precipitate directly from seawater, in close association with heterotrophic oxidation of organic matter (Chow and Goldberg, 1960). Indeed, the abundance of particulate Ba and presumably barite is found to peak in the upper mesopelagic zone, where maximum regeneration of organic matter occurs (e.g., Sternberg et al., 2008). Living phytoplankton contain a relatively large amount of labile Ba, which is released rapidly on remineralization and may provide the main source of Ba for barite precipitation in the microenvironments. The process is possibly mediated by marine bacteria that produce extracellular polymeric substances that bind Ba and serve as crystal nucleation sites (González-Muñoz et al., 2012; Martinez-Ruiz et al., 2018). The particulate pelagic barite export flux is thus related to the integrated C export below the euphotic zone (i.e., export production – the amount of organic matter produced in the ocean by primary production that is not recycled [remineralized] before it sinks into the

Figure 1 Pelagic barite crystals from sediment core Leg 199, Site 1221C, Core 11–3, 66–70 cm depth. Sample separated following the sequential leaching process of Eagle et al. (2003).

aphotic zone) and is not necessarily associated with specific organisms (e.g., Jaquet et al., 2007).

Assuming the mechanisms of pelagic barite formation and preservation in the marine environment at present have also operated in the past, quantitative relations between excess Ba (Ba_{ex}, Ba that is not of terrigenous origin; Dymond et al., 1992; Averyt and Paytan 2004) or pelagic barite (BaAR) accumulation in marine sediments (Paytan et al., 1996a) can be used to reconstruct export production (Figure 2). Indeed, owing to the relatively high preservation of barite in non-sulfate-reducing sediments (Paytan and Kastner, 1996) and the general non-species-specific relation to export production, Ba_{ex} or BaAR has been widely used to reconstruct ocean export production over different geological time scales (Schmitz, 1987; Rutsch et al., 1995; Paytan et al., 1996a; Dean et al., 1997; Nürnberg et al., 1997; Bonn et al., 1998; Bains et al., 2000; González-Muñoz et al., 2003; Averyt and Paytan, 2004; Jaccard et al., 2005; Olivarez Lyle and Lyle, 2006; Griffith et al., 2010; Erhardt et al., 2013; Ma et al., 2014, 2015; Carter et al., 2016). When using BaAR to quantitatively reconstruct export production, care must be taken to use accumulation rates and not the abundance of the mineral because of the impact of dilution by other sedimentary phases. Moreover, the pelagic nature (i.e., formation in the water column in association with sinking particulate matter, as opposed to other formation mechanisms; see Section 2) of the barite crystals must be confirmed using microscopy or isotopic analyses (Paytan et al., 2002; Griffith et al., 2018).

2 Nonpelagic Marine Barite

For reconstructing ocean export production, nonpelagic marine barite (i.e., hydrothermal, cold-seep, and diagenetic barite) deposits should be avoided, as they do not represent water column processes. Such samples can be generally

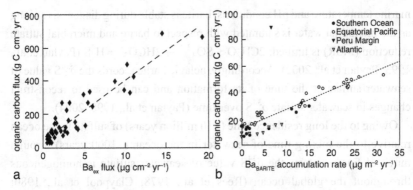

Figure 2 (a) Relation between excess Ba and organic carbon export in sediment traps (modified from Dymond et al., 1992). (b) Relation between barite Ba (Ba$_{BARITE}$) accumulation rate in core-top sediments and carbon export in the overlying water column (modified from Eagle et al., 2003). Note the outlier in, which is one sample from the highly productive coastal Peru Margin, was not included in the regression and may not be representative of the Holocene period.

averted with careful site selection to avoid (1) sites that are close to hydrothermal and cold seeps settings and (2) areas with extensive sulfate reduction (based on sulfate concentrations in porewaters), since pelagic barite in these sediments could be dissolved under sulfate-reducing conditions that deplete sulfate in porewater fluids and lower the barite saturation state. The released Ba can then diffuse upward and subsequently reprecipitate as diagenetic barite on encountering residual sulfate that is isotopically enriched. Because the process of precipitation of these barite deposits differs from that occurring in the upper water column, the crystal size and habit of these barite crystals and their chemical and isotopic composition differ from those of the pelagic barite crystals that form in the water column in association with organic matter regeneration (Paytan et al., 2002; Griffith et al., 2018). Accordingly, barite samples should be screened visually (with a scanning electron microscope) and/or chemically to verify their water column pelagic origin.

3 Sulfur Isotopes

Pelagic barite incorporates sulfate into the crystal structure with little fractionation of sulfur isotopes (<0.4‰; Paytan et al., 1998). The sulfur isotopic ratio (δ^{34}S) of core-top pelagic barite is consistent with the seawater sulfate δ^{34}S value of 21‰ in the present-day open ocean (Rees et al., 1978; Paytan et al., 1998; Markovic et al., 2015). Owing to the low solubility of barite at marine temperature and pH conditions, barite is stable during diagenesis as long as the interstitial water is saturated with respect to barite and microbial sulfate reduction (MSR) is limited: $2CH_2O + SO_4^{2-} \rightarrow 2HCO_3^- + H_2S$ (Paytan et al., 1993; Paytan et al., 2002). Accordingly, pelagic barite records the $\delta^{34}S$ value of seawater sulfate at the time of its formation and can be used to reconstruct changes in seawater sulfate $\delta^{34}S$ over time (Paytan et al., 1998, 2004).

Owing to the long residence time (>10 million years) of sulfate in the ocean relative to the mixing time of seawater in the ocean (~1000 years) through much of Earth history, the $\delta^{34}S$ value of seawater sulfate is homogeneous throughout the global ocean (Rees et al., 1978; Claypool et al., 1980; Jørgensen and Kasten, 2006). The $\delta^{34}S$ value of seawater sulfate is controlled by a balance between S inputs via terrestrial weathering and volcanic degassing and S outputs via burial of sulfur-bearing minerals (e.g., pyrite, gypsum) and their respective isotopic compositions. The process of MSR preferentially incorporates ^{32}S in the sulfide produced, a fraction of which reacts with reactive iron and precipitates as pyrite. As a result of the large isotopic fractionation associated with MSR, pyrite sulfur is isotopically light relative to seawater sulfate. Thus, precipitation of sedimentary pyrite often exerts dominant control on seawater sulfate $\delta^{34}S$ values because the residual sulfate becomes enriched in ^{34}S as more pyrite is buried. By contrast, sulfate-bearing minerals (e.g., gypsum, anhydrite) precipitation involves little or no fractionation relative to coeval seawater. The $\delta^{34}S$ value of weathered evaporites thus varies spatially as a function of the geological age of the weathered terrain. Sulfate sourced from volcanism and hydrothermal activity is also important; these sources typically have a $\delta^{34}S$ value of approximately 0‰ (relative to Vienna Canyon Diablo Troilite [VCDT]; Sakai et al., 1982). Consequently, fluctuations in the $\delta^{34}S$ value of seawater sulfate through time (Figure 3) can shed light on potential changes in geological, geochemical, and biological processes that affect the sources and sinks of S to/from the ocean and their isotopic values (e.g., Paytan et al., 1998, 2004; Wortmann and Paytan, 2012; Markovic et al., 2015; Yao et al., 2018).

The advantage of using pelagic barite to reconstruct paleoseawater sulfate $\delta^{34}S$ signatures is an improved temporal resolution and reduced uncertainty associated with precipitation in marginal environments and potential diagenetic alteration of reconstructions using evaporite deposits that have intermittent occurrence and poor age control (Claypool et al., 1980; Strauss, 1997). Paytan et al. (1998, 2004) presented a continuous pelagic barite $\delta^{34}S$ curve for the past 130 million years at a temporal resolution of less than 1 million years. The barite record has a relatively narrow range of oceanic sulfate $\delta^{34}S$ values for each age (<0.5‰). During times of rapid perturbations (e.g.,

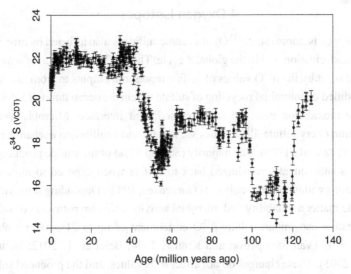

Figure 3 Seawater sulfate S isotopic curve based on pelagic barite for the last 130 million years. Error bars are 1σ. Data sources: Paytan et al. (1998); Markovic et al. (2015, 2016); Masterson et al. (2016); Yao et al. (2018, 2020).

the Paleocene–Eocene Thermal Maximum, Eocene, Holocene), higher temporal resolution records (a few thousand years) provide precise information of the magnitude, duration, and gradient of the $\delta^{34}S$ excursions, permitting the use $\delta^{34}S$ for stratigraphic correlations (e.g., Markovic et al., 2015; Yao et al., 2018).

Barite that precipitates from other sources of sulfate (i.e., hydrothermal or porewater fluids) may have different S-isotope signatures than those of seawater sulfate. Diagenetic barite precipitates from porewater fluids where sulfate is depleted by MSR, leaving residual sulfate highly enriched $\delta^{34}S$ values (up to 84‰; Torres et al., 1996; Paytan et al., 2002); barite that precipitates in these porewaters hence assumes these enriched isotopic values. In anaerobic sediments of cold-seep environments, sulfate is also the electron acceptor driving anaerobic oxidation of methane (AOM): $CH_4 + SO_4^{2-} \rightarrow HCO_3^- + HS^- + H_2O$ (Jørgensen and Kasten, 2006). Cold-seep barite $\delta^{34}S$ values represent a mixture of seawater and isotopically enriched porewater sulfate that has undergone AOM (Greinert et al., 2002; Paytan et al., 2002). Hydrothermal barite that forms in association with volcanic hydrothermal activity can be isotopically depleted relative to coeval seawater sulfate $\delta^{34}S$ because of the incorporation of various amounts of S from hydrothermal fluids (Paytan et al., 2002).

4 Oxygen Isotopes

The oxygen isotopic ratio ($\delta^{18}O$) of oceanic sulfate is also affected by processes and transformation within the global S cycle. The $\delta^{18}O$ composition of seawater sulfate is set by the $\delta^{18}O$ values of sulfate inputs and outputs to/from the ocean as modified by microbial recycling of sulfate within the ocean through MSR and sulfide reoxidation near the seawater–sediment interface. Microbial sulfate reduction alters sulfate $\delta^{18}O$ values via kinetic and equilibrium exchange reactions (Fritz et al., 1989). The majority (80% to 95%) of the sulfide produced by MSR is subsequently reoxidized back to sulfate when exposed to more oxic conditions within marine sediments (Jørgensen, 1982). Depending primarily on organic matter availability and microbial activity, different pathways of sulfide reoxidation can impart sulfate $\delta^{18}O$ enrichment of up to 21‰ over ambient water $\delta^{18}O$ (Van Stempvoort and Krouse, 1994; Balci et al., 2012; Böttcher et al., 2005). These changes do not affect $\delta^{34}S$ values, and the produced sulfate retains the S-isotope composition of the sulfide that was oxidized. Because of the rapid O-isotope exchange during microbial sulfur cycling, the residence time of sulfate-bound oxygen is on the order of 1 million years (Jørgensen and Kasten, 2006; Markovic et al., 2016).

Pelagic barite has been used as an archive for the reconstruction of the $\delta^{18}O$ composition of oceanic sulfate (Turchyn and Schrag, 2004, 2006; Markovic et al., 2016). The $\delta^{18}O$ value of core-top pelagic barite is about 1.4‰ to 2.5‰ depleted relative to the present-day seawater sulfate $\delta^{18}O$ value of 8.6‰ Vienna Standard Mean Ocean Water (VSMOW). This isotopic offset is likely a result of a kinetic effect associated with barite precipitation (Turchyn and Schrag., 2004) and/or organic sulfur oxidation during barite precipitation (Markovic et al., 2016). While the underlying factors controlling this isotopic offset require further investigation, if we assume that the offset is constant through time, it is possible to use barite $\delta^{18}O$ to shed light on fluctuation in the $\delta^{18}O$ values of seawater sulfate and the processes that control these values. Turchyn and Schrag (2004, 2006) reported oceanic sulfate $\delta^{18}O$ data derived from pelagic barite over the Cenozoic (Figure 4) and showed distinct trends in the temporal evolution of sulfate $\delta^{18}O$ and $\delta^{34}S$ values. It has been suggested that the temporal variation in oceanic sulfate $\delta^{18}O$ is linked to changes in the aerial distribution of continental shelves with organic-rich sediments, which in turn corresponds to sea-level variations (Turchyn and Schrag, 2006; Markovic et al., 2016).

Oxygen–isotope exchange between sulfate and seawater is strongly dependent on the temperature and pH of the solution (Chiba and Sakai, 1985). In the present-day ocean, the oxygen–isotope exchange rate is very slow and does not

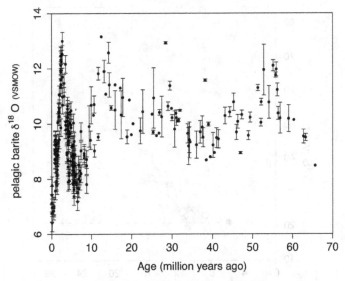

Figure 4 Pelagic barite O isotopic curve for the Cenozoic. Error bars are 1σ.
Data sources: Turchyn and Schrag (2004, 2006).

approach equilibrium within the estimated residence time of sulfate in the ocean (Chiba and Sakai, 1985). Thus, seawater sulfate $\delta^{18}O$ is out of isotopic equilibrium with oceanic water (0‰ VSMOW; Van Stempvoort and Krouse, 1994), and this is also reflected in pelagic barite. By contrast, hydrothermal barite forms under high temperatures (e.g., >120°C) and near-neutral pH conditions, recording sulfate $\delta^{18}O$ that is in equilibrium with ambient hydrothermal fluid $\delta^{18}O$ (Chiba and Sakai, 1985; Alt et al., 2010). For example, the $\delta^{18}O$ value of hydrothermal barite from Juan de Fuca Ridge black smoker varies between 3‰ and 16‰ (Goodfellow et al., 1993). Higher values (e.g., 20.6‰) were reported for ancient hydrothermal barite deposits, although these samples may have been affected by postdepositional alteration or metamorphism (e.g., Moles et al., 2014). The O-isotope fractionation factor of the sulfate–water system can also be used as a geothermometer to estimate the temperature of geothermal reservoirs (Chiba and Sakai, 1985).

Cold-seep barite and diagenetic barite are usually isotopically enriched in ^{18}O resulting from O-isotope fractionation during sulfate reduction, a process that preferentially reduces sulfate with depleted O-isotope values leaving any residual sulfate isotopically enriched (Torres et al., 1996; Greinert et al., 2002). Although the parameters controlling O-isotope fractionation between sulfate and water are not fully known, it is now understood that the kinetic fractionation associated with MSR for oxygen isotopes is up to 25% of that for sulfur isotopes

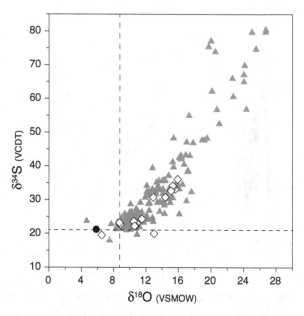

Figure 5 Sulfur and oxygen isotopes of barite formed in different environments. Black dashed lines denote the values of modern seawater sulfate. The black circle is modern pelagic barite; open diamonds are hydrothermal barite; and gray triangles are diagenetic and cold-seep barite. Figure after Griffith et al. (2018) using data sources: Aquilina et al. (1997); Greinert et al. (2002); Paytan et al. (2002); Kim et al. (2004); Feng and Roberts (2011); Eickmann et al. (2014); Stevens et al. (2015); Markovic et al. (2016); Griffith et al. (2018).

and depends on the cell-specific sulfate reduction rate (Aharon and Fu, 2000), while O-isotope exchange reactions drive the residual sulfate $\delta^{18}O$ toward an equilibrium value of ~29‰ more elevated than the ambient water $\delta^{18}O$ at ~5°C (Fritz et al., 1989). Investigation of combined $\delta^{18}O$ and $\delta^{34}S$ signatures recorded in barite can be used to identify the environment where barite precipitates and forms (Griffith and Paytan, 2012; Griffith et al., 2018) (Figure 5).

5 Strontium Isotopes

Because of the similarity between the alkali earth metals strontium (Sr^{2+}) and barium (Ba^{2+}), Sr^{2+} can substitute for Ba^{2+} in barite at concentrations >10,000 ppm (i.e., >1%, or >10,000 μg Sr per g barite; Averyt and Paytan, 2003). Rubidium (Rb^+) content is low in barite, eliminating any need for correcting for in situ production of radiogenic ^{87}Sr from the decay of ^{87}Rb. Thus, pelagic barite serves as a reliable archive of the contemporaneous radiogenic Sr isotopic composition of seawater ($^{87}Sr/^{86}Sr$) (Goldberg et al., 1969; Paytan et al., 1993).

Owing to the long residence time of Sr in seawater (~2.5 million years; Hodell et al., 1990) compared to the mixing time of the ocean, the Sr isotopic composition of seawater is uniform throughout the ocean. Changes in the seawater $^{87}Sr/^{86}Sr$ in the past recorded in pelagic barite or other archives (e.g., carbonate, apatite) are thought to reflect changes in weathering and hydrothermal activity, related to changes in climate and tectonics (e.g., Burke et al., 1982; Palmer and Edmond, 1989; Hodell et al., 1990; McArthur et al., 2012). The seawater radiogenic Sr-isotope curve also serves to aid in stratigraphic correlation and dating, and its record in barite is particularly useful for dating carbonate-poor or diagenetically altered sedimentary sections where sulfate reduction is not dominating the system (Mearon et al., 2003).

Pelagic barite also serves as an archive of the stable (nonradiogenic) Sr isotopic composition ($\delta^{88/86}Sr$) of seawater (Griffith et al., 2018). However, pelagic barite $\delta^{88/86}Sr$ values are depleted relative to dissolved Sr in seawater, likely because of both equilibrium and chemical kinetic effects during crystal growth (Widanagamage et al., 2014). Unlike other types of barite (e.g., hydrothermal., continental), pelagic barite has a consistent offset of −0.53‰ from the seawater $\delta^{88/86}Sr$ value as determined by measurements of core-top samples (Griffith et al., 2018). Assuming that this offset is constant through time and independent of environmental conditions, the seawater $\delta^{88/86}Sr$ can be reconstructed using pelagic barite (Paytan et al., 2017). Marine carbonate is the major sink for Sr, which undergoes mass-dependent isotopic fractionation during carbonate precipitation (Krabbenhöft et al., 2010). Consequently, stable Sr isotopes have been suggested as a proxy for reconstructing carbonate deposition in the ocean through time (Krabbenhöft et al., 2010; Vollstaedt et al., 2014; Pearce et al., 2015). By combining records of seawater $\delta^{88/86}Sr$ and $^{87}Sr/^{86}Sr$, it should be possible to simultaneously reconstruct changes in continental weathering and global carbonate burial (e.g., Vollstaedt et al., 2014; Pearce et al., 2015).

6 Calcium Isotopes

Calcium (Ca^{2+}) is an alkali earth metal that can substitute for Ba^{2+} in barite similarly to Sr^{2+}, however, to a much smaller degree (~300 ppm or 300 μg Ca per g barite; Averyt and Paytan, 2003). Calcium has six stable isotopes that exhibit mass-dependent isotopic fractionation between the dissolved Ca^{2+} in the ocean and carbonate ($CaCO_3$) minerals formation and deposition – the major sink of Ca from the ocean. Therefore, similar to the stable Sr isotopic system ($\delta^{88/86}Sr$), changes in the Ca isotopic composition of the ocean ($\delta^{44/40}Ca$) could be used to reconstruct carbonate burial and dissolution – although other changes in the marine Ca cycle such as the continental weathering flux, alteration of oceanic crust, and the hydrothermal flux may also play an important role in changes in seawater $\delta^{44/40}Ca$ (Skulan et al., 1997; De La Rocha and DePaolo, 2000; Fantle and DePaolo, 2005; Heuser et al., 2005; Farkas et al., 2007; Blättler and Higgins, 2017).

The Ca isotopic composition of pelagic barite from core-top samples shows a consistent isotopic fractionation from seawater that does not appear to be dependent on any environmental parameter measured (e.g., temperature, salinity, water column barite saturation; Griffith et al., 2008b). Assuming that this isotopic fractionation is constant through time, pelagic barite can be used as a "passive" tracer of seawater $\delta^{44/40}Ca$ (i.e., decoupled from the global output flux of Ca; Fantle, 2010). The use of pelagic barite also eliminates complications related to biological fractionation associated with biogenic calcium carbonate precipitation, which, as stated earlier, serves as the major sink of Ca from the ocean (Skulan et al., 1997; De La Rocha and DePaolo, 2000; Fantle and DePaolo, 2005; Heuser et al., 2005; Sime et al., 2005; Hippler et al., 2006; Farkas et al., 2007). By combining records of pelagic barite and carbonate, both the seawater $\delta^{44/40}Ca$ and the fractionation factor associated with carbonate sedimentation can be defined through time to gain insight into changes in the cycling of Ca (and C) in the ocean (Figure 6; Griffith et al., 2008a, 2011; Fantle, 2010; Fantle and Tipper, 2014).

Even a highly stable sulfate mineral such as barite can be susceptible to postdepositional change because of diagenetic reactions that alter its stable isotopic composition (e.g., ocean acidification at the Paleocene–Eocene Thermal Maximum; Griffith et al., 2015). Stable Ca isotopes appear to be sensitive to these diagenetic processes occurring during extreme climatic or environmental perturbations, whereas radiogenic Sr ratios are not (Griffith et al., 2015). Within the upper part of the sedimentary column, radiogenic Sr in the porewater has been shown to remain dominated by seawater, even where carbonate is rapidly dissolving (e.g., Fantle, 2015). However, these diagenetic effects need to be taken into account when interpreting Ca isotope data, and possibly other geochemical proxies over extreme climatic events that drive sediment dissolution and alteration. Multiple sites and different sedimentary phases (e.g., pelagic barite, carbonate, bioapatite) are needed to reconstruct global changes in seawater chemistry, and in particular Ca isotopes, reliably.

7 Barium Isotopes

With the advancement of multicollector inductively coupled plasma mass spectrometry (MC-ICP-MS), barium isotopes in diverse Earth materials,

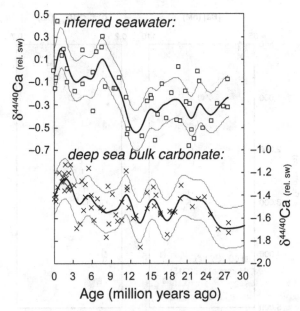

Figure 6 Inferred seawater Ca isotopic curve derived from pelagic barite (open squares) and deep-sea bulk carbonate (Ca sink; x) relative to modern seawater with smoothed cubic spline fits of the data (dark lines) with ± 0.18‰ (average $2\sigma_{mean}$) in gray following Griffith et al. (2008a). Data sources: Griffith et al. (2008a); Fantle and DePaolo (2005, 2007).

including barite, have been determined. During barite precipitation the light Ba isotopes are incorporated preferentially into the barite crystal structure with fractionation of ~0.5‰ relative to seawater (von Allmen et al., 2010; Horner et al., 2015). The precipitation of barite in the marine water column appears to be the primary mediator of Ba isotopic fractionation in the open ocean (Horner et al., 2015). The effects of this isotopic fractionation cause a distinct depth profile structure of Ba isotopes in the ocean (Figure 7) whereby the surface ocean that is depleted in Ba is characterized by enrichment in Ba isotopic compositions ($\delta^{138/134}Ba_{NIST} \sim$ +0.6‰), whereas deep waters have higher Ba concentrations and $\delta^{138/134}Ba_{NIST} \sim$ +0.3‰. Barite in deep-sea marine environments has a Ba isotopic composition close to $\delta^{138/134}Ba_{NIST} \sim$0‰ (Bridgestock et al., 2018). Current research is focused on understanding the systematics and utility of Ba isotopes in barite specifically to refine the understanding of past changes in the global Ba cycle, ocean circulation, and productivity. As the number of laboratories that have the capability of Ba isotopic analyses increase, we expect that the use of this relatively new proxy will expand.

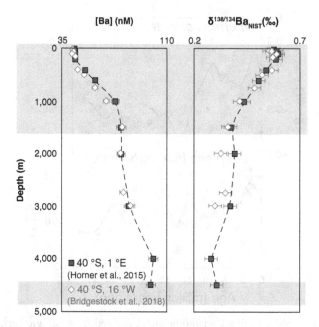

Figure 7 Dissolved Ba concentrations and Ba isotopic ratios in seawater. Green squares and blue open diamonds denote data from the South Atlantic after Horner et al. (2015) and Bridgestock et al. (2018), respectively.

8 Future Prospects

Barite accumulation rates in marine sediments and the stable isotopic compositions of the major constituents (S, O, Sr, Ca, and Ba) in pelagic barite have added valuable insights that have fundamentally enhanced understanding of ocean productivity and chemistry over a range of timescales. However, the abundance and isotopic composition of minor constituents in barite have yet to be systematically explored. For example, pelagic barite may record intermediate-water-depth radiogenic isotopes (e.g., neodymium) that have shorter residence times than the global oceanic mixing and can be used as a tracer of intermediate-water-mass paleocirculation (e.g., Martin et al., 1995). The decay of radium, thorium, and lead in pelagic barite can be utilized to estimate sedimentation rates and absolute ages of sediments for the Holocene (e.g., Paytan et al., 1996b; van Beek et al., 2001). The ability to analyze additional tracers in barite will open new insights and opportunities for understanding present and past biogeochemical cycles and their relation to Earth processes.

References

Key References

Dehairs E., Stroobants N., and Goeyens L. (1991) Suspended barite as a tracer of biological activity in the Southern Ocean. *Mar. Chem.* **35**, 399–410.

In this article the distribution of particulate barite suspended in seawater is described showing a subsurface maximum associated with the depth of organic matter regeneration in the water column and the oxygen minimum zone.

Dymond J., Suess E., and Lyle M. (1992) Barium in deep-sea sediments: A geochemical proxy for paleoproductivity. *Paleoceanography* **7**, 163–181.

The first quantitative use of biogenic barium to reconstruct export productivity. This is based on relationship between Ba and organic C in sediment trap material and core top sediments.

Ganeshram R. S., Francois R., Commeau J., and Brown-Leger S. L. (2003) An experimental investigation of barite formation in seawater. *Geochim. Cosmochim. Acta* **67**, 2599–2605.

An experimental mesocosm setup that simulated barite production in seawater from decaying organic matter from phytoplankton cultures. This shows that barite forms regardless of the specific organism used.

Griffith E. M., and Paytan A. (2012) Barite in the ocean: Occurrence, geochemistry and palaeoceangraphic applications. *Sedimentology* **59**, 1817–1835.

A review article summarizing the formation and use of barite in marine sediments for palaeoceanographic applications.

Griffith E. M., Paytan A., Caldeira K., Bullen T. D., and Thomas E. (2008a) A dynamic marine calcium cycle during the past 28 million years. *Science* **322**, 1671–1674.

The first report on Ca isotope analyses in pelagic barite and their fluctuations over the past 28 million years.

Griffith E. M., Paytan A., Wortmann U. G., Eisenhauer A., and Scher H. D. (2018b) Combining metal and nonmetal isotopic measurements in barite to identify mode of formation. *Chem. Geol.* **500**, 148–158.

In this article the authors investigate the isotopic signatures of barite that is formed in different terrestrial and marine settings and use the differences to distinguish modes of formation.

Horner T. J., Kinsley C. W., and Nielsen S. G. (2015) Barium isotopic fractionation in seawater mediated by barite cycling and oceanic circulation. *Earth Planet. Sci. Lett.* **430**, 511–522.

Barium isotopes of dissolved and particulate Ba in seawater are reported and the distribution linked to formation and dissolution of barite in the marine water column.

Martinez-Ruiz F., Jroundi F., Paytan A., Guerra-Tschuschke I., Abad M. M., and González-Muñoz M.T. (2018) Barium bioaccumulation by bacterial biofilms and implications for Ba cycling and use of Ba proxies. *Nat. Commun.* **9**, 1619.

Barite is formed in seawater media by bacterial mediation and the particulate barite crystals that are formed are investigated under the scanning electron microscope to reveal nucleation on phosphorus-rich biofilms.

Monnin C., Jeandel C., Cattaldo T., and Dehairs F. (1999) The marine barite saturation state of the world's oceans. *Mar. Chem.* **65**, 253–261.

A thorough investigation of the concentrations on Ba in seawater and the saturation state of seawater with respect to barite. The authors demonstrate that much of the seawater is undersaturated with respect to barite.

Paytan A., and Griffith E. M. (2007) Marine barite: Recorder of variations in ocean export productivity. *Deep Sea Res. Part II* **54**, 687–705.

A review on the formation of barite in seawater and its utility as a proxy for export productivity.

Paytan A., Kastner M., Martin E. E., Macdougall J. D., and Herbert T. (1993) Marine barite as a monitor of seawater strontium isotope composition. *Nature* **366**, 445–449.

The first report on Sr isotopes in pelagic barite over the past 35 million years clearly demonstrating that barite incorporates and preserves the seawater Sr isotopic composition.

Paytan A., Kastner M., Campbell D., and Thiemens M. H. (2004) Seawater sulfur isotope fluctuations in the Cretaceous. *Science* **304**, 1663–1665.

A high-resolution seawater sulfate S isotope curve over the past 130 million years derived from pelagic barite and an interpretation of the causes of the observed fluctuations.

Turchyn A. V., and Schrag D. P. (2004) Oxygen isotope constraints on the sulfur cycle over the past 10 million years. *Science* **303**, 2004–2007.

A high-resolution seawater sulfate O isotope curve over the past 10 million years derived from pelagic barite and an interpretation of the causes of the observed fluctuations.

Supporting References

Aharon P., and Fu B. (2000) Microbial sulfate reduction rates and sulfur oxygen isotope fractionations at oil and gas seeps in deepwater Gulf of Mexico. *Geochim. Cosmochim. Acta* **64**, 233–246.

Alt J. C., Laverne C., Coggon R. M., et al. (2010) Subsurface structure of a submarine hydrothermal system in ocean crust formed at the East Pacific Rise, ODP/IODP Site 1256. *Geochem. Geophy. Geosy.* **11**, Q10010.

Aquilina L., Dia A. N., Boulegue J., Bourgois J., and Fouillac A. M. (1997) Massive barite deposits in the convergent margin off Peru: Implications for fluid circulation. *Geochim. Cosmochim. Acta* **61**, 1233–1245.

Averyt K. B., and Paytan A. (2003) Empirical partition coefficients for Sr and Ca in marine barite: Implications for reconstructing seawater Sr and Ca concentrations. *Geochem. Geophy. Geosy.* **4**, 1043.

Averyt K. B., and Paytan A. (2004) A comparison of multiple proxies for export production in the equatorial Pacific. *Paleoceanography* **19**, PA4003.

Bains S., Norris R. D., Corfield R. M., and Faul K. L. (2000) Termination of global warmth at the Palaeocene/Eocene boundary through productivity feedback. *Nature* **407**, 171–174.

Balci N., Mayer B., Shanks W. C., and Mandernack K. W. (2012) Oxygen and sulfur isotope systematics of sulfate produced during abiotic and bacterial oxidation of sphalerite and elemental sulfur. *Geochim. Cosmochim. Acta* **77**, 335–351.

Bernstein R. E., and Byrne R. H. (2004) Acantharians and marine barite. *Mar. Chem.* **86**, 45–50.

Bishop J. K. B. (1988) The barite-opal-organic carbon association in oceanic particulate matter. *Nature* **332**, 341–343.

Blättler C. L., and Higgins J. A. (2017) Testing Urey's carbonate-silicate cycle using the calcium isotopic composition of sedimentary carbonates. *Earth Planet. Sci. Lett.* **479**, 241–251.

Bonn W. J., and Gingele F. X. (1998) Palaeoproductivity at the Antarctic continental margin: Opal and barium records for the last 400 ka. *Palaeogeogr. Palaeoclimatol. Palaeoecol.* **139**, 195–211.

Böttcher M.E., Thamdrup B., Gehre M., and Theune A. (2005) $^{34}S/^{32}S$ and $^{18}O/^{16}O$ fractionation during sulfur disproportionation by *Desulfobulbus propionicus*. *Geochem. J.* **22**, 219–226.

Bridgestock L., Hsieh Y. T., Porcelli D., Homoky W. B., Bryan A., and Henderson, G. M. (2018) Controls on the barium isotope compositions of marine sediments. *Earth Planet. Sci. Lett.* **481**, 101–10.

Burke W. H., Denison R. E., Hetherington E. A., Koepnick R. B., Nelson H. F., and Otto J. B. (1982) Variation of sea-water $^{87}Sr/^{86}Sr$ throughout Phanerozoic time. *Geology* **10**, 516–519.

Carter S. C., Griffith E. M., and Penman D. E. (2016) Peak intervals of equatorial Pacific export production during the middle Miocene climate transition. *Geology* **44**, 923–926.

Chiba H., and Sakai H. (1985) Oxygen Isotope exchange-rate between dissolved sulfate and water at hydrothermal temperatures. *Geochim. Cosmochim. Acta* **49**, 993–1000.

Chow T. J., and Goldberg E.D. (1960) On the marine geochemistry of barium. *Geochim. Cosmochim. Acta* **20**, 192–198.

Claypool G. E., Holser W. T., Kaplan I. R., Sakai H., and Zak I. (1980) The age curves of sulfur and oxygen isotopes in marine sulfate and their mutual interpretation. *Chem. Geol.* **28**, 199–260.

Dean W. E., Gardner J. V., and Piper D. A. (1997) Inorganic geochemical indicators of glacial-interglacial changes in productivity and anoxia on the California continental margin. *Geochim. Cosmochim. Acta* **61**, 4507–4518.

Dehairs F., Chesselet R., and Jedwab J. (1980) Discrete suspended particles of barite and the barium cycle in the open ocean. *Earth Planet. Sci. Lett.* **49**, 528–550.

Dehairs E., Fagel N., Antia A. N., Peinert R., Elskens M., and Goeyens L. (2000) Export production in the Bay of Biscay as estimated from barium-barite in settling material: A comparison with new production. *Deep Sea Res. Part I* **47**, 583–601.

De La Rocha L., and DePaolo D. J. (2000) Isotopic evidence for variations in the marine calcium cycle over the Cenozoic. *Science* **289**, 1176–1178.

Dymond J., and Collier R. (1996) Particulate barium fluxes and their relationships to biological productivity. *Deep Sea Res. Part II* **43**, 1283–1308.

Eagle M., Paytan A., Arrigo K. R., van Dijken G., and Murray R. W. (2003) A comparison between excess barium and barite as indicators of carbon export. *Paleoceanography* **18**, 21.1–21.13.

Eickmann B., Thorseth I. H., Peters M., Strauss H., Brocker M., and Pedersen R. B. (2014) Barite in hydrothermal environments as a recorder of subseafloor processes: A multiple-isotope study from the Loki's Castle vent field. *Geobiology* **12**, 308–321.

Erhardt A. M., Pälike H., and Paytan A. (2013) High-resolution record of export production in the eastern equatorial Pacific across the Eocene-Oligocene transition and relationships to global climatic records. *Paleoceanography* **28**, 130–142.

Esser B. K., and Volpe A. M. (2002) At-sea high-resolution chemical mapping: Extreme barium depletion in North Pacific surface water. *Mar. Chem.* **79**, 67–79.

Fantle M. S. (2010) Evaluating the Ca isotope proxy. *Am J Sci.* **310**, 194–230.

Fantle M. S. (2015) Calcium isotopic evidence for rapid recrystallization of bulk marine carbonates and implications for geochemical proxies. *Geochim Cosmochim Acta* **148**, 378–401.

Fantle M. S., and DePaolo D. J. (2005) Variations in the marine Ca cycle over the past 20 million years. *Earth Planet. Sci. Lett.* **237**, 102–117.

Fantle M. S., and DePaolo D. J. (2007) Ca isotopes in carbonate sediment and pore fluid from ODP 807a: The Ca^{2+}(aq)-calcite equilibrium fractionation factor and calcite recrystallization rates in Pleistocene sediments. *Geochem. Cosmochim. Acta* **71**, 2524–2546.

Fantle M. S., and Tipper E. T. (2014) Calcium isotopes in the global biogeochemical Ca cycle: Implications for development of a Ca isotope proxy. *Earth-Sci. Rev.* **129**, 148–177.

Farkas J., Buhl D., Blenkinsop J., and Veizer J. (2007) Evolution of the oceanic calcium cycle during the late Mesozoic: Evidence from $\delta^{44/40}Ca$ of marine skeletal carbonates. *Earth Planet. Sci. Lett.* **253**, 96–111.

Feng D., and Roberts H. 2011. Geochemical characteristics of the barite deposits at cold seeps from the northern Gulf of Mexico. *Earth Planet. Sci. Lett.* **309**, 89–99.

Finlay B. J., Hetherington N. B., and Davison W. (1983) Active biological participation in lacustrine barium chemistry. *Geochim. Cosmochim. Acta* **47**, 1325–1329.

Francois R., Honjo S., Manganini S. J., and Ravizza G. E. (1995) Biogenic barium fluxes to the deep: Implications for paleoproductivity reconstruction. *Global Biogeochem. Cy.* **9**, 289–303.

Fritz P., Basharmal G. M., Drimmie R. J., Ibsen J., and Qureshi R. M. (1989) Oxygen isotope exchange between sulphate and water during bacterial reduction of sulphate. *Chem. Geol.* **79**, 99–105.

Goldberg E. D., Somayajulu B. L. K., Gallway J., Kaplann I. R., and Faure G. (1969) Difference between barites of marine and continental origins. *Geochim. Cosmochim. Acta* **33**, 287–289.

González-Muñoz M. T., Fernandez-Luque B., Martinez-Ruiz F., et al. (2003) Precipitation of barite by *Myxococcus xanthus*: Possible implications for the biogeochemical cycle of barium. *Appl. Environ. Microbiol.* **69**, 5722–5725.

González-Muñoz M. T., Martinez-Ruiz F., Morcillo F., Martin-Ramos J. D., and Paytan A. (2012) Precipitation of barite by marine bacteria: A possible mechanism for marine barite formation. *Geology* **40**, 675–678.

Gooday A. J., and Nott J. A. (1982) Intracellular barite crystals in two Xenophyaphores, *Aschenonella ramuliformia* and *Galatheammina* sp. with comments on the taxonomy of A. Ramuliformia. *J. Mar. Biol. Assoc. UK* **62**, 595–605.

Goodfellow W. D., Grapes K., Cameron B., and Franklin J. M. (1993) Hydrothermal alteration associated with massive sulfide deposits, middle valley, northern Juan de Fuca ridge. *Can. Mineral.* **31**, 1025–1060.

Greinert J., Bollwerk S. M., Derkachev A., Bohrmann G., and Suess E. (2002) Massive barite deposits and carbonate mineralization in the Derugin Basin, Sea of Okhotsk: Precipitation processes at cold seep sites. *Earth Planet. Sci. Lett.* **203**, 165–180.

Griffith E.M., Calhoun M., Thomas E., et al.(2010) Export productivity and carbonate accumulation in the Pacific Basin at the transition from a greenhouse to icehouse climate (late Eocene to early Oligocene). *Paleoceanography* **25**, PA3212.

Griffith E. M., Fantle M., Eisenhauer A., Paytan A., and Bullen T.D. (2015) Effects of ocean acidification on the marine calcium isotope record at the Paleocene-Eocene Thermal Maximum. *Earth Planet. Sci. Lett.* **419**, 81–92.

Griffith E. M., Paytan A., Eisenhauer A., Bullen T. D., and Thomas E. (2011) Seawater calcium isotope ratios across the Eocene-Oligocene transition. *Geology* **39**, 683–686.

Griffith E. M., Schauble E. A., Bullen T. D., and Paytan A. (2008b) Characterization of calcium isotopes in natural and synthetic barite. *Geochim. Cosmochim. Acta* **72**, 5641–5658.

Heuser A., Eisenhauer A., Böhm F., et al. (2005) Calcium isotope ($\delta^{44/40}$Ca) variations of Neogene planktonic foraminifera. *Paleoceanography* **20**, PA2013.

Hippler D., Eisenhauer A., and Nagler T. F. (2006) Tropical Atlantic SST history inferred from Ca isotope thermometry over the last 140 ka. *Geochim. Cosmochim. Acta* **70**, 90–100.

Hodell D. A., Mead G. A., and Mueller P. A. (1990) Variation in the strontium isotopic composition of seawater (8 Ma to present): Implications for chemical weathering rates and dissolved fluxes to the oceans. *Chem. Geol.* **80**, 291–307.

Jaccard S. L., Haug G. H., Sigman D. M., Pedersen T. F., Thierstein H. R., and Röhl U. (2005) Glacial/interglacial changes in subarctic North Pacific stratification. *Science* **308**, 1003–1006.

Jaquet S. H. M., Dehairs F., Elskens M., Savoye N., and Cardinal D. (2007) Barium cycling along WOCE SR3 line in the Southern Ocean. *Mar. Chem.* **106**, 33–45.

Jørgensen B. B. (1982) Mineralization of organic-matter in the sea bed – the role of sulfate reduction. *Nature* **296**, 643–645.

Jørgensen B. B., and Kasten S. (2006) Sulfur cycling and methane oxidation. In H. D. Schulz and M. Zabel (eds.),. *Marine Geochemistry*, 2nd ed., pp. 271–309.Berlin: Springer.

Kim J., Lee I., and Lee K.-Y. (2004) S, Sr, and Pb isotopic systematics of hydrothermal chimney precipitates from the Eastern Manus Basin, western

Pacific: Evaluation of magmatic contribution to hydrothermal system. *J. Geophys. Res.* **109**, B12210.

Krabbenhöft A., Eisenhauer A., Böhm F., et al.(2010) Constraining the marine strontium budget with natural strontium isotope fractionations (^{87}Sr/^{86}Sr*, $\delta^{88/86}$Sr) of carbonates, hydrothermal solutions and river water. *Geochim. Cosmochim. Acta* **74**, 4097–4109.

Ma Z., Gray E., Thomas E., Murphy B., Zachos J., and Paytan A. (2014). Carbon sequestration during the Palaeocene-Eocene Thermal Maximum by an efficient biological pump. *Nat. Geosci.* **7**, 382–388.

Ma Z., Ravelo A. C., Liu Z., Zhou L., and Paytan A. (2015) Export production fluctuations in the eastern equatorial Pacific during Pliocene-Pleistocene: Reconstruction using barite accumulation rates. *Paleoceanogr. Paleocl.* **30**, 1455–1469.

Markovic S., Paytan A., Li H., and Wortmann U. G. (2016) A revised seawater sulfate oxygen isotope record for the last 4 Myr. *Geochim. Cosmochim. Acta* **175**, 239–251.

Markovic S., Paytan A., and Wortmann U. G. (2015) Pleistocene sediment offloading and the global sulfur cycle. *Biogeoscience* **12**, 3043–3060.

Martin E. E., Macdougall J. D., Herbert T. D., Paytan A., and Kastner M. (1995) Strontium and neodymium isotopic analyses of marine barite separates. *Geochim. Cosmochim. Acta* **59**, 1353–1361.

Masterson A. L., Wing B. A., Paytan A., Farquhar J., and Johnston D. T. (2016) The minor sulfur isotope composition of Cretaceous and Cenozoic seawater sulfate. *Paleoceanography* **31**, 779–788.

McArthur J. M., Howarth R. J., and Shields G. A. (2012) Strontium isotope stratigraphy. In F. M. Gradstein, J. G. Ogg, M. Schmitz, and G. Ogg (eds.), *The Geologic Time Scale 2012*, pp. 127–144. Philadelphia: Elsevier.

Mearon S., Paytan A., and Bralower T. J. (2003) Cretaceous strontium isotope stratigraphy using marine barite. *Geology* **31**, 15–18.

Moles N. R., Boyce A. J., and Fallick A. E. (2014) Abundant sulphate in the Neoproterozoic ocean: Implications of constant δ^{34}S barite in the Aberfeldy SEDEX deposits, Scottish Dalradian. In G. R. T. Jenkin, P. A. J. Lusty, I. McDonald, M. P. Smith, A. J. Boyce, and J. J. Wilkinson (eds.), *Ore Deposits in an Evolving Earth*, Vol. 393, pp. 189–212. Special Publication. London: Geological Society.

Monnin C., and Cividini D. (2006) The saturation state of the world's ocean with respect to (Ba, Sr)SO$_4$ solid solutions. *Geochim. Cosmochim. Acta* **70**, 3290–3298.

Nürnberg C. C., Bohrmann G., and Schlüter M. (1997) Barium accumulation in the Atlantic sector of the Southern Ocean: Results from 190,000-year record. *Paleoceanography* **12**, 574–603.

Olivarez Lyle A., and Lyle M. W. (2006) Missing organic carbon in Eocene marine sediments: Is the metabolism the biological feedback that maintains end-member climates? *Paleoceanogr. Paleocl.* **21**, PA2007.

Palmer M. R., and Edmond J. M. (1989) The strontium isotope budget of the modern ocean. *Earth Planet. Sci. Lett.* **92**, 11–26.

Paytan A., Eisenhauer A., Wallmann K. J. G., Griffith E. M., and Ridgwell A. (2017) Stable and radiogenic Sr isotopes in barite – Clues on the links between weathering, climate and the C cycle (invited). *EOS Trans. AGU*, Fall Meet. Suppl., Abstract PP14A-01.

Paytan A., and Kastner M. (1996) Benthic Ba fluxes in the central Equatorial Pacific, implications for the oceanic Ba cycle. *Earth Planet. Sci. Lett.* **142**, 439–450.

Paytan, A., Kastner M., Campbell D., and Thiemens M. H. (1998) Sulfur isotopic composition of Cenozoic seawater sulfate. *Science* **282**, 1459–1462.

Paytan A., Kastner M., and Chavez F. P. (1996a) Glacial to interglacial fluctuations in productivity in the Equatorial Pacific as indicated by marine barite. *Science* **274**, 1355–1357.

Paytan A., Mearon S., Cobb K., and Kastner M. (2002) Origin of marine barite deposits: Sr and S isotope characterization. *Geology* **30**, 747–750.

Paytan A., Moore W. S., and Kastner M. (1996b) Sedimentation rate as determined by ^{266}Ra activity in marine barite. *Geochim. Cosmochim. Acta* **60**, 4131–4319.

Pearce C. R., Parkinson I. J., Gaillardet J., Charlier B. L. A., Mokadem F., and Burton K. W. (2015) Reassessing the stable ($\delta^{88/86}$Sr) and radiogenic ($\delta^{87/86}$Sr) strontium isotopic composition of marine inputs. *Geochim. Cosmochim. Acta* **157**, 125–146.

Rees C. E., Jenkins W. J., and Monster J. (1978) The sulphur isotopic composition of ocean water sulphate. *Geochim. Cosmochim. Acta* **42**, 377–381.

Rushdi A., McManus J., and Collier R. (2000) Marine barite and celestite saturation in seawater. *Mar. Chem.* **69**, 19–31.

Rutsch H. J., Mangini A., Bonai G., Dittrich-Hannen B., Kubik P. W., Suter M., and Segl M. (1995) ^{10}Be and Ba concentrations in West African sediments trace productivity in the past. *Earth Planet. Sci. Lett.* **133**, 129–143.

Sakai H., Casadevall T. J., and Moore J. G. (1982) Chemistry and isotope ratios of sulfur in basalts and volcanic gases at Kilauea Volcano, Hawaii. *Geochim. Cosmochim. Acta* **46**, 729–738.

Schmitz B. (1987) Barium, equatorial high productivity, and the wandering of the Indian continent. *Paleoceanography* **2**, 63–77.

Sime N. G., De La Rocha C. L., and Galy A. (2005) Negligible temperature dependence of calcium isotope fractionation in 12 species of planktonic foraminifera. *Earth Planet. Sci. Lett.* **232**, 51–66.

Skulan J., DePaolo D. J., and Owens T. L. (1997) Biological control of calcium isotopic abundances in the global calcium cycle. *Geochim. Cosmochim. Acta* **61**, 2505–2510.

Sternberg E., Jeandel C., Robin E., and Souhaut N. (2008) Seasonal cycle of suspended barite in the Mediterranean Sea. *Geochim. Cosmochim. Acta* **72**, 4020–4034.

Stevens E. W. N., Bailey J. V., Flood B. E., et al. (2015) Barite encrustation of benthic sulfur-oxidizing bacteria at a marine cold seep. *Geobiology* **13**, 588–603.

Strauss H. (1997) The isotopic composition of sedimentary sulfur through time. *Palaeogeogr. Palaeoclimatol. Paleoecol.* **132**, 97–118.

Torres M. E., Brumsack H. J., Bohrmann G., and Emeis K. C. (1996) Barite fronts in continental margin sediments: A new look at barium remobilization in the zone of sulfate reduction and formation of heavy barites in authigenic fronts. *Chem. Geol.* **127**, 125–139.

Turchyn A. V., and Schrag D. P. (2006) Cenozoic evolution of the sulfur cycle: Insight from oxygen isotopes in marine sulfate. *Earth Planet. Sci. Lett.* **241**, 763–779.

van Beek P., and Reyss J. L. (2001) ^{226}Ra in marine barite: New constraints on supported ^{226}Ra. *Earth Planet. Sci. Lett.* **187**, 147–161.

Van Stempvoort D. R., and Krouse H. R. (1994) Controls of d18O in sulfate: Review of experimental data and application to specific environments. In N. Alpers and D. W. Blowes (eds.), *Environmental Geochemistry of Sulfide Oxidation*, Vol. 550, pp. 446–480. Washington, DC: American Chemical Society.

Vollstaedt H., Eisenhauer A., Wallann K., et al. (2014) The Phanerozoic $\delta^{88/86}$Sr record of seawater: New constrains on past changes in oceanic carbonate fluxes. *Geochim. Cosmochim. Acta* **128**, 249–265.

von Allmen K., Böttcher M. E., Samankassou E., and Nägler T. F. (2010) Barium isotope fractionation in the global barium cycle: First evidence from barium minerals and precipitation experiments. *Chem. Geol.*, **277**, 70–77.

Widanagamage I. H., Schauble E. A., Scher H. D., and Griffith E. M. (2014) Stable Sr isotope fractionation in synthetic barite. *Geochim. Cosmochim. Acta* **147**, 58–74.

Wortmann U. G., and Paytan A. (2012) Rapid variability of seawater chemistry over the past 130 million years. *Science* **337**, 334–336.

Yao W., Paytan A., Griffith E. M., Martinez-Ruiz F., Markovic S., and Wortmann U.G. (2020) A revised seawater sulfate S-isotope curve for the Eocene. *Chem. Geol.* **532**, 119382.

Yao W., Paytan A., and Wortmann U. G. (2018) Large-scale ocean deoxygenation during the Paleocene-Eocene Thermal Maximum. *Science* **361**, 804–806.

Cambridge Elements ≡

Elements in Geochemical Tracers in Earth System Science

Timothy Lyons
University of California

Timothy Lyons is a Distinguished Professor of Biogeochemistry in the Department of Earth Sciences at the University of California, Riverside. He is an expert in the use of geochemical tracers for applications in astrobiology, geobiology and Earth history. Professor Lyons leads the 'Alternative Earths' team of the NASA Astrobiology Institute and the Alternative Earths Astrobiology Center at UC Riverside.

Alexandra Turchyn
University of Cambridge

Alexandra Turchyn is a University Reader in Biogeochemistry in the Department of Earth Sciences at the University of Cambridge. Her primary research interests are in isotope geochemistry and the application of geochemistry to interrogate modern and past environments.

Chris Reinhard
Georgia Institute of Technology

Chris Reinhard is an Assistant Professor in the Department of Earth and Atmospheric Sciences at the Georgia Institute of Technology. His research focuses on biogeochemistry and paleoclimatology, and he is an Institutional PI on the 'Alternative Earths' team of the NASA Astrobiology Institute.

About the Series

This innovative series provides authoritative, concise overviews of the many novel isotope and elemental systems that can be used as 'proxies' or 'geochemical tracers' to reconstruct past environments over thousands to millions to billions of years – from the evolving chemistry of the atmosphere and oceans to their cause-and-effect relationships with life.

Covering a wide variety of geochemical tracers, the series reviews each method in terms of the geochemical underpinnings, the promises and pitfalls, and the 'state-of-the-art' and future prospects, providing a dynamic reference resource for graduate students, researchers and scientists in geochemistry, astrobiology, paleontology, paleoceanography and paleoclimatology.

The short, timely, broadly accessible papers provide much-needed primers for a wide audience – highlighting the cutting-edge of both new and established proxies as applied to diverse questions about Earth system evolution over wide-ranging time scales.

Elements in Geochemical Tracers in Earth System Science

Printed in the United States
By Bookmasters